El Niño
What Is This?

By

Buchi Nwadiuto

DEDICATION

This book is dedicated to all victims of El- Niño

PREFACE

El Niño is defined by prolonged warming in the Pacific Ocean sea surface temperatures when compared with the average value. The U.S NOAA definition is a 3-month average warming of at least 0.5 °C (0.9 °F) in a specific area of the east-central tropical Pacific Ocean; other organizations define the term slightly differently. Typically, this anomaly happens at irregular intervals of two to seven years, and lasts nine months to two years. The average period length is five years. When this warming occurs for seven to nine months, it is classified as El Niño "conditions"; when its duration is longer, it is classified as an El Niño "episode. When El Niño conditions last for many months, extensive ocean warming and the reduction in easterly trade winds limits upwelling of cold nutrient-rich deep water, and its economic impact to local fishing for an international market can be serious.

More generally, El Niño can affect commodity prices and the macro-economy of different countries. It can constrain the supply of rain-driven agricultural commodities; reduce agricultural output, construction, and services activities; create food-price and generalized inflation; and may trigger social unrest in commodity-dependent poor countries that primarily rely on imported food

Table of Contents

CHAPTER ONE

DEFINITION OF EL NIÑO---------------------------------

CHAPTER TWO

WHAT IS ENSO AND WHAT DOES IT STANDS FOR? ---

CHAPTER THREE

HOW DID THESE TERMS ORIGINATE---------------

CHAPTER FOUR

EFFECTS OF ENSO WARM PHASE (EL NIÑO) ------

CHAPTER FIVE

HEALTH AND SOCIAL IMPACTS-----------------------

CHAPTER SIX

SOUTH AMERICA--

CHAPTER SEVEN

NORTH AMERICA---

CHAPTER EIGHT

HOW DO EL NIÑO AND LA NIÑA AFFECT WEATHER IN OTHER PARTS OF THE WORLD-----

CHAPTER NINE

DOES EL NIÑO AFFECT GLOBAL TEMPERATURE---

CHAPTER TEN

RELATION TO CLIMATE CHANGE--------------------

CHAPTER ELEVEN

EL NIÑO DIVERSITY--

CHAPTER TWELVE

HAS EL NIÑO ABANDONED L.A? ----------------------

CHAPTER THIRTEEN

WHAT FORCES CAUSE EL NIÑO? ---------------------

CHAPTER ONE
DEFINITION OF EL NIÑO

El Niño /ɛl ˈniːnjoʊ/ (Spanish pronunciation: [el ˈniɲo])
is the warm phase of the El Niño Southern Oscillation
(commonly called ENSO) and is associated with a band
of warm ocean water that develops in the central and
east-central equatorial Pacific (between approximately
the International Date Line and 120°W), including off
the Pacific coast of South America. El Niño Southern
Oscillation refers to the cycle of warm and cold
temperatures, as measured by sea surface temperature,
SST, of the tropical central and eastern Pacific Ocean. El
Niño is accompanied by high air pressure in the western
Pacific and low air pressure in the eastern Pacific. The
cool phase of ENSO is called "La Niña" with SST in the
eastern Pacific below average and air pressures high in
the eastern and low in western Pacific. The ENSO cycle,
both El Niño and La Niña, causes global changes of both
temperatures and rainfall. Mechanisms that cause the
oscillation remain under study. El Niño (Spanish name
for the male child), initially referred to a weak, warm
current appearing annually around Christmas time along
the coast of Ecuador and Peru and lasting only a few
weeks to a month or more. Every three to seven years,
an El Niño event may last for many months, having
significant economic and atmospheric consequences
worldwide. During the past forty years, ten of these
major El Niño events have been recorded, the worst of
which occurred in 1997-1998. Previous to this, the El
Niño event in 1982-1983 was the strongest. Some of the
El Niño events have persisted more than one year.

In the tropical Pacific, trade winds generally drive the surface waters westward. The surface water becomes progressively warmer going westward because of its longer exposure to solar heating. El Niño is observed when the easterly trade winds weaken, allowing warmer waters of the western Pacific to migrate eastward and eventually reach the South American Coast. The cool nutrient-rich sea water normally found along the coast of Peru is replaced by warmer water depleted of nutrients, resulting in a dramatic reduction in marine fish and plant life.

Developing countries dependent upon agriculture and fishing, particularly those bordering the Pacific Ocean, are the most affected. In Spanish, the capitalized term "El Niño" refers to the Child Jesus, so named because the pool of warm water in the Pacific near South America is often at its warmest around Christmas. "La Niña", chosen as the 'opposite' of El Niño, literally translates to "the girl child".

Measurements and simulations have found that climate change has created a tendency toward more extreme El Niños in recent years. El Niño is defined by prolonged warming in the Pacific Ocean sea surface temperatures when compared with the average value. The U.S NOAA definition is a 3-month average warming of at least 0.5 °C (0.9 °F) in a specific area of the east-central tropical Pacific Ocean; other organizations define the term slightly differently. Typically, this anomaly happens at irregular intervals of two to seven years, and lasts nine months to two years. The average period length is five years. When this warming occurs for seven to nine months, it is classified as El Niño "conditions"; when its duration is longer, it is classified as an El Niño "episode".

The first signs of an El Niño are a weakening of the Walker circulation or trade winds and strengthening of the Hadley circulation and may include:

Rise in surface pressure over the Indian Ocean, Indonesia, and Australia;

Fall in air pressure over Tahiti and the rest of the central and eastern Pacific Ocean;

Trade winds in the south Pacific weaken or head east;

Warm air rises near Peru, causing rain in the northern Peruvian deserts

El Niño's warm rush of nutrient-poor water heated by its eastward passage in the Equatorial Current, replaces the cold, nutrient-rich surface water of the Humboldt Current.

A recent study has appeared applying network theory to the analysis of El Niño events; the study presented evidence that the dynamics of a described "climate network" were very sensitive to such events, with many links in the network failing during the events.

CHAPTER TWO
WHAT IS ENSO AND WHAT DOES IT STANDS FOR?

El Niño and La Niña are normally accompanied by variations in the tropical Pacific Ocean's Walker Circulation, as well as a vast see-saw in atmospheric pressure—the Southern Oscillation—that modifies the Walker pattern. The term El Niño–Southern Oscillation, or ENSO, refers to the combination of atmospheric and oceanic effects associated with both El Niño and La Niña.

El Niño and La Niña can be thought of as the ocean part of ENSO, while the Southern Oscillation can be thought of as the atmospheric component. Through the atmosphere, tele-connections occur via waves to influence regions far from the tropical Pacific. ENSO is a coupled phenomenon that would not occur without interactions between the atmosphere and ocean. Sometimes an atmospheric shift can occur in the tropical Pacific without the ocean fully responding, or vice versa. However, for an El Niño or La Niña to develop, the atmosphere and ocean must evolve in sync. For example, the usual east-to-west surface trade winds across the tropical Pacific tend to weaken during El Niño; this allows warm water to shift toward the eastern tropical Pacific. That warm water supports the eastward development of showers and thunderstorms, and the resulting atmospheric circulations tend to weaken the trade winds even further. Such positive feedbacks are crucial for the development of both El Niño and La Niña events.

ENSO is one of the main sources of year-to-year variability in weather and climate around the world. Research to more fully incorporate this inter-annual variability into computer models is a major focus for improving long-range forecasting.

CHAPTER THREE
HOW DID THESE TERMS ORIGINATE?

The name El Niño originated in the region where one of the phenomenon's local effects was first recognized. Fishing people and other coastal residents of Peru had long noticed a strange feature in the eastern Pacific Ocean waters that border their home. This region of tropical yet relatively cool water is host to one of the world's most productive fisheries and a large bird population. In the first months of each year, a warm southward current usually modified the cool waters. But every few years, this warming started early (in December), was far stronger, and lasted as long as a year or two. Torrential rains fell on the arid land; as one early observer put it, "the desert becomes a garden." Warm waters flowing south brought water snakes, bananas, and coconuts from equatorial rain forests. However, the same current shut off the deeper, cooler waters that are crucial to sustaining the region's marine life.

Because these strong events were often observed close to Christmas time, the phenomenon was dubbed El Niño (when capitalized, "the little boy" becomes "the Christ Child" in Spanish). Although "la niña" refers to a young girl, the capitalized term La Niña did not originate in the same way; it was adopted by researchers in the 1970s and 1980s to illustrate the relationship between warming and cooling events in the waters of the eastern tropical Pacific. Some researchers suggested "El Viejo" (the old man) or "anti–El Niño" as alternatives, but La Niña has won out as the standard term.

CHAPTER FOUR
EFFECTS OF ENSO WARM PHASE (EL NIÑO)

When El Niño conditions last for many months, extensive ocean warming and the reduction in easterly trade winds limits upwelling of cold nutrient-rich deep water, and its economic impact to local fishing for an international market can be serious.

More generally, El Niño can affect commodity prices and the macro economy of different countries. It can constrain the supply of rain-driven agricultural commodities; reduce agricultural output, construction, and services activities; create food-price and generalized inflation; and may trigger social unrest in commodity-dependent poor countries that primarily rely on imported food. A University of Cambridge Working Paper shows that while Australia, Chile, Indonesia, India, Japan, New Zealand and South Africa face a short-lived fall in economic activity in response to an El Niño shock, other countries may actually benefit from an El Niño weather shock (either directly or indirectly through positive spillovers from major trading partners), for instance, Argentina, Canada, Mexico and the United States. Furthermore, most countries experience short-run inflationary pressures following an El Niño shock, while global energy and non-fuel commodity prices increase. The IMF estimates a significant El Niño can boost the GDP of the United States by about 0.5% (due largely to lower heating bills) and reduce the GDP of Indonesia by about 1.0%.

CHAPTER FIVE
HEALTH AND SOCIAL IMPACTS

Extreme weather conditions related to the El Niño cycle correlate with changes in the incidence of epidemic diseases. For example, the El Niño cycle is associated with increased risks of some of the diseases transmitted by mosquitoes, such as malaria, dengue, and Rift Valley fever. Cycles of malaria in India, Venezuela, Brazil, and Colombia have now been linked to El Niño. Outbreaks of another mosquito-transmitted disease, Australian encephalitis (Murray Valley encephalitis—MVE), occur in temperate south-east Australia after heavy rainfall and flooding, which are associated with La Niña events. A severe outbreak of Rift Valley fever occurred after extreme rainfall in north-eastern Kenya and southern Somalia during the 1997–98 El Niño.

ENSO conditions have also been related to Kawasaki disease incidence in Japan and the west coast of the

United States, via the linkage to tropospheric winds across the North Pacific Ocean.

ENSO may be linked to civil conflicts. Scientists at The Earth Institute of Columbia University, having analyzed data from 1950 to 2004, suggest ENSO may have had a role in 21% of all civil conflicts since 1950, with the risk of annual civil conflict doubling from 3% to 6% in countries affected by ENSO during El Niño years relative to La Niña years.

CHAPTER SIX
SOUTH AMERICA

Because El Niño's warm pool feeds thunderstorms above, it creates increased rainfall across the east-central and eastern Pacific Ocean, including several portions of the South American west coast. The effects of El Niño in South America are direct and stronger than in North America. An El Niño is associated with warm and very wet weather months in April–October along the coasts of northern Peru and Ecuador, causing major flooding whenever the event is strong or extreme. The effects during the months of February, March, and April may become critical. Along the west coast of South America, El Niño reduces the upwelling of cold, nutrient-rich water that sustains large fish populations, which in turn sustain abundant sea birds, whose droppings support the fertilizer industry. The reduction in upwelling leads to fish kills off the shore of Peru.

The local fishing industry along the affected coastline can suffer during long-lasting El Niño events. The world's largest fishery collapsed due to overfishing during the 1972 El Niño Peruvian anchoveta reduction. During the 1982–83 event, jack mackereland anchoveta populations were reduced, scallops increased in warmer water, but hake followed cooler water down the continental slope, while shrimp and sardines moved southward, so some catches decreased while others increased.Horse mackerel have increased in the region during warm events. Shifting locations and types of fish due to changing conditions provide challenges for fishing industries. Peruvian sardines have moved during El Niño events to Chilean areas. Other conditions provide further complications, such as the government of Chile in 1991 creating restrictions on the fishing areas for self-employed fishermen and industrial fleets.

The ENSO variability may contribute to the great success of small, fast-growing species along the Peruvian coast, as periods of low population removes predators in the area. Similar effects benefit migratory birds that travel each spring from predator-rich tropical areas to distant winter-stressed nesting areas.

Southern Brazil and northern Argentina also experience wetter than normal conditions, but mainly during the spring and early summer. Central Chile receives a mild winter with large rainfall, and the Peruvian-Bolivian Altiplano is sometimes exposed to unusual winter snowfall events. Drier and hotter weather occurs in parts of the Amazon River Basin, Colombia, and Central America

CHAPTER SEVEN
NORTH AMERICA

Winters, during the El Niño effect, are warmer and drier than average in the Northwest, northern Midwest, and upper Northeast United States, so those regions experience reduced snowfalls. Meanwhile, significantly wetter winters are present in northwest Mexico and the southwest United States, including central and southern California, while both cooler and wetter than average winters in northeast Mexico and the Southeastern United States (including the Tidewater region of Virginia) occur during the El Niño phase of the oscillation.

Some believed the ice storm in January 1998, which devastated parts of New England, southern Ontario and southern Quebec, was caused or accentuated by El Niño's warming effects. However, according to Amir Shabbar, an El Niño expert from Environment Canada, there is no unequivocal link, although it could have been a contributing factor.

El Niño warmed Vancouver for the 2010 Winter Olympics, such that the area experienced a warmer than average winter during the games.

The synoptic condition for the Tehuantepecer is associated with high-pressure system forming in Sierra Madre of Mexico in the wake of an advancing cold front, which causes winds to accelerate through the Isthmus of Tehuantepec. Tehuantepecers primarily occur during the cold season months for the region in the wake of cold fronts, between October and February, with a summer maximum in July caused by the westward extension of the Azores High. Wind magnitude is greater during El Niño years than during La Niña years, due to the more frequent cold frontal incursions during El Niño winters. Its effects can last from a few hours to six days. El Niño is credited with suppressing Atlantic hurricanes, and made the 2009 Atlantic hurricane season the least active in 12 years.

CHAPTER EIGHT
HOW DO EL NIÑO AND LA NIÑA AFFECT WEATHER IN OTHER PARTS OF THE WORLD

During an El Niño event, as warm surface water in the Pacific is shifted thousands of miles east of its usual location, the showers and thunderstorms nurtured by convection above this warm tropical water also change location. As the rising motion associated with the convection also shifts eastward, it causes other adjustments in atmospheric circulation, well away from the tropical Pacific. These persistent zones of rising and sinking air can shift into new locations for months, causing prolonged wet or dry conditions and related atmospheric heating anomalies. In turn, the anomalous heating sets up planetary-scale waves in the atmosphere that radiate away from the region, especially into the hemisphere experiencing winter. These are "teleconnections"—large-scale, long-lasting shifts in atmospheric circulation that can affect much of the globe. The effects extend throughout the Pacific Rim, across large parts of North America, and on to eastern Africa and other regions. La Niña brings a different set of teleconnections to these and other regions, with some but not all effects roughly opposite to those of El Niño.

The most common teleconnections associated with El Niño and La Niña during northern summer and winter. Not every warm- or cool-water event will produce all of these impacts, because other atmospheric features interact with each ENSO event to influence weather and climate around the globe. Weaker events, in particular, may look quite different from the prototypes shown here. El Niño Modoki events (where the warming is concentrated further to the west than usual) show significant differences in teleconnections from other El Niño events.

In the United States, a strong El Niño event tends to produce milder- and drier-than-average conditions toward the north and cooler- and wetter-than-average conditions to the south. In California, a strong El Niño very often brings more moisture than usual. However, during the weakest El Niño events, San Francisco and Los Angeles are a bit more likely to be unusually dry than unusually wet. (Meteorologist Jan Null maintains a compilation of additional El Niño "myths and realities.")

In Africa, East Africa — including Kenya, Tanzania, and the White Nile basin — experiences, in the long rains from March to May, wetter-than-normal conditions. Conditions are also drier than normal from December to February in south-central Africa, mainly in Zambia, Zimbabwe, Mozambique, and Botswana. Direct effects of El Niño resulting in drier conditions occur in parts of Southeast Asia and Northern Australia, increasing bush fires, worsening haze, and decreasing air quality dramatically. Drier-than-normal conditions are also in general observed in Queensland, inland Victoria, inland New South Wales, and eastern Tasmania from June to August.

Many ENSO linkages exist in the high southern latitudes around Antarctica. Specifically, El Niño conditions result in high pressure anomalies over the Amundsen and Bellingshausen Seas, causing reduced sea ice and increased pole ward heat fluxes in these sectors, as well as the Ross Sea. The Weddell Sea, conversely, tends to become colder with more sea ice during El Niño. The exact opposite heating and atmospheric pressure anomalies occur during La Niña. This pattern of variability is known as the Antarctic dipole mode, although the Antarctic response to ENSO forcing is not ubiquitous.

El Niño's effects on Europe appear to be strongest in winter. Recent evidence indicates that El Niño causes a colder, drier winter in Northern Europe and a milder, wetter winter in Southern Europe. The El Niño winter of 2009/10 was extremely cold in Northern Europe but El Niño is not the only factor at play in European winter weather and the weak El Niño winter of 2006/2007 was unusually mild in Europe, and the Alps recorded very little snow coverage that season.

As warm water spreads from the west Pacific and the Indian Ocean to the east Pacific, it takes the rain with it, causing extensive drought in the western Pacific and rainfall in the normally dry eastern Pacific. Singapore experienced the driest February in 2014 since records began in 1869, with only 6.3 mm of rain falling in the month and temperatures hitting as high as 35 °C on 26 February. The years 1968 and 2005 had the next driest Februaries, when 8.4 mm of rain fell.

CHAPTER NINE
DOES EL NIÑO AFFECT GLOBAL TEMPERATURE

During El Niño, a deep pool of warm water usually restricted to the western tropical Pacific is replaced by a much larger, shallower pool of warm water that covers most or the entire tropical Pacific. The expanded zone of warm sea surface temperatures allows more heat to be conveyed from the ocean into the atmosphere for months at a time. As a result, globally averaged temperatures often rise by a few tenths of a degree Fahrenheit during the latter stages of a strong El Niño event. Conversely, global temperatures can drop by a similar amount during a La Niña event.

NCAR scientist Kevin Trenberth has likened El Niño to a "pressure valve" that releases built-up heat from the oceans into the atmosphere. The oceans cool during El Niño events, while the global atmosphere warms. Scientists often account for ENSO by factoring out these bumps and dips in global temperature when analyzing the long-term trends related to climate change. For example, the first 15 years of this century saw more La Niña than El Niño influence, and global air temperatures showed little rise. Prior to that period, in the 1980s and 1990s, when El Niño events were more frequent, global temperatures rose more sharply.

Note that factors also influence global temperature, such as the eruption of Mt. Pinatubo in 1991. The volcano threw enough sun-blocking material into the atmosphere to cause a drop in global temperatures during 1992, despite the presence of El Niño.

Shifts in global temperature, as well as in the likelihood of ENSO events, are closely associated with the state of the Pacific Decadal Oscillation (PDO), a pattern of ocean temperatures that reverses every 20-30 years. More La Niña events tend to be observed when the PDO is negative, and more El Niño events when it is positive. Scientists are not yet sure what prompts the PDO to shift modes.

CHAPTER TEN
RELATION TO CLIMATE CHANGE

Despite steady improvement, it remains difficult for many global climate models to simulate all aspects of ENSO events, so it has been a challenge to determine whether or how El Niño and La Niña will be affected by rising global temperatures. Some models have suggested that the central Pacific could warm enough to produce El Niño–like conditions on a semi-permanent basis. The most recent (2013) assessment by the Intergovernmental Panel on Climate Change concluded:

There is high confidence that the El Niño-Southern Oscillation (ENSO) will remain the dominant mode of inter annual variability in the tropical Pacific, with global effects in the 21st century. Due to the increase in moisture availability, ENSO-related precipitation variability on regional scales will likely intensify. Natural variations of the amplitude and spatial pattern of ENSO are large and thus confidence in any specific projected change in ENSO and related regional phenomena for the 21st century remains low.

During the last several decades the number of El Niño events increased, although a much longer period of observation is needed to detect robust changes. The question is, or was, whether this is a random fluctuation or a normal instance of variation for that phenomenon or the result of global climate changes as a result of global warming. A 2014 study reported a robust tendency to more frequent extreme El Niños, occurring in agreement with a separate recent model prediction for the future.

Several studies of historical data suggest the recent El Niño variation is linked to anthropogenic climate change; in accordance with the larger consensus on climate change. For example, even after subtracting the positive influence of decade-to-decade variation (which is shown to be present in the ENSO trend), the amplitude of the ENSO variability in the observed data still increases, by as much as 60% in the last 50 years.

It may be that the observed phenomenon of more frequent and stronger El Niño events occurs only in the initial phase of the climate change, and then (e.g., after the lower layers of the ocean get warmer, as well), El Niño will become weaker than it was. It may also be that the stabilizing and destabilizing forces influencing the phenomenon will eventually compensate for each other. More research is needed to provide a better answer to that question. However, a new 2014 model appearing in a research report indicated unmitigated climate change would particularly affect the surface waters of the eastern equatorial Pacific and possibly double extreme El Niño occurrences.

CHAPTER ELEVEN
EL NIÑO DIVERSITY

The traditional Niño, also called Eastern Pacific (EP) El Niño, involves temperature anomalies in the Eastern Pacific. However, in the last two decades, nontraditional El Niños were observed, in which the usual place of the temperature anomaly (Niño 1 and 2) is not affected, but an anomaly arises in the central Pacific (Niño 3.4). The phenomenon is called Central Pacific (CP) El Niño,[63] "dateline" El Niño (because the anomaly arises near the dateline), or El Niño "Modoki" (Modoki is Japanese for "similar, but different"). There are flavors of ENSO additional to EP and CP types and some scientists argue that ENSO exists as a continuum often with hybrid types.

The effects of the CP El Niño are different from those of the traditional EP El Niño—e.g., the recently discovered El Niño leads to more hurricanes more frequently making landfall in the Atlantic.

The recent discovery of El Niño Modoki has some scientists believing it to be linked to global warming. However, comprehensive satellite data go back only to

1979. More research must be done to find the correlation and study past El Niño episodes. More generally, there is no scientific consensus on how/if climate change may affect ENSO.

There is also a scientific debate on the very existence of this "new" ENSO. Indeed, a number of studies dispute the reality of this statistical distinction or its increasing occurrence, or both, either arguing the reliable record is too short to detect such a distinction, finding no distinction or trend using other statistical approaches, or that other types should be distinguished, such as standard and extreme ENSO The first recorded El Niño that originated in the central Pacific and moved toward the east was in 1986. Recent Central Pacific El Niños happened in 1986–87, 1991–92, 1994–95, 2002–03, 2004–05 and 2009–10. Furthermore, there were "Modoki" events in 1957–59, 1963–64, 1965–66, 1968–70, 1977–78 and 1979–80.

CHAPTER TWELVE
HAS EL NIÑO ABANDONED L.A?

By this point in winter, Southern California was supposed to be dealing with rains and flooding, not brush fires and beach weather.

Yet temperatures have soared this week, breaking records in downtown Los Angeles and other locations across California, with even hotter conditions expected Tuesday. Forecasters warn of more hot winds as well as temperatures that could exceed 90 degrees downtown.

For all the talk of monster rains from El Niño, all but three days in the last month have been dry in the Los Angeles area.

It's too early to be certain. But some scientists say El Niño is operating differently than they expected — at least for Southern California.

In the fall, the consensus was that El Niño would give Southern California the best chance for above-average rains and much less of a chance in Northern California. But the opposite has turned out to be true.

Southern California is still well below average rainfall, with downtown L.A. reporting 52% of normal since Oct. 1. But deluge after deluge to the north has built back the snowpack — it's 105% of normal in the Sierra Nevada — and begun to refill drought-depleted reservoirs.

For Southern California, the strong El Niño "hasn't been a great predictor so far this winter," said Stanford University climate scientist Daniel Swain, and "hasn't been influencing the atmosphere in exactly the same way that we have seen."

A massive ridge of high pressure is keeping much of California dry and warm this week. In Southern California, that brought dry winds and temperatures that approached 90 degrees.

At Mt. Baldy, the heat prompted officials to close main trails after unusually icy conditions resulted in two deaths and numerous helicopter rescues.

"The warm weather is melting the snow and it freezes at night," said Sherry Rollman, a spokeswoman for Angeles National Forest.

The conditions led to 12 people being airlifted off the mountain over one weekend.

A 47-year-old man died recently when he slipped and fell off the side of Icehouse Saddle near Mt. Baldy. Dong Xing Liu, who was known as Tony Liu, was the second person to die in the area last week. One Tuesday, Daniel Nguyen, 23, slipped and fell 1,500 feet after struggling to help a friend on the Devil's Backbone trail. The San Bernardino County Sheriff's Department said it took nearly three hours in ice, snow and wind to retrieve his body.

Signs along the trails advise hikers to use equipment such as crampons and ice axes. It was not immediately clear what type of gear Liu, Nguyen and the rescued hikers were using, but officials said at least a handful were not wearing crampons — spiked metal plates affixed to boots to make it easier to walk on ice — and didn't have ice axes.

Mt. Baldy Fire Department Capt. Gordon Greene said a few people rescued one Saturday afternoon were wearing shoe chains, which are less effective than crampons.

A series of big El Niño-influenced storms the first week of January left Mt. Baldy and other winter resorts with snow. But the promised "conveyor belt" of storms has not materialized since then.

El Niño is the warming of ocean waters about 1,000 to 2,000 miles south of California, along the equator.

That water heats up, fueling thunderstorms that push warm air into the atmosphere, which travels north. Eventually, it falls back down to the ocean in the subtropics, at roughly the same latitude Hawaii sits, Swain said.

It's that movement in the atmosphere — a circulation pattern called the "Hadley cell" — that supercharges a subtropical jet stream from Japan eastward into Southern California and into the southern United States.

But something changed this year. With the zone of warm water in the ocean particularly large and persistent, the movement of warm air above it traveled farther north than expected. That means the parade of storms zipping across the Pacific Ocean established a path over Northern California and even the Pacific Northwest — and bypassed Southern California, Swain said.

That may be the reason why all but one storm have missed L.A. over the last month. "It may be because El Niño is so strong," Swain said.

The difference in the path of the jet stream is "very slight in the global context," Swain said in an interview. "But if you're in Los Angeles, that difference means a lot."

Across L.A, the heat wave brought longing for the promised rain that hasn't materialized.

Gloria Lopez, a Boyle Heights resident, held her 5-year-old daughter tightly by the hand as they walked along 1st Street. While Lopez, 48, was wearing sandals, her daughter was a little less prepared in a long-sleeved shirt.

"I don't know what's happening with this climate," Lopez said. "They said El Niño was coming, but nothing has happened. I don't know what's going to happen."

"El calor no se quiere ir," she said. "The heat doesn't want to go away."

Setrak Malatian, a Pasadena resident, was annoyed by the sun.

"Wintertime should be winter, not wintertime should be summer," he said.

Some remained optimistic about the possibility of rain down the line.

"When we get nice weather, then I'm happy for it," said Toni Fields, who was dressed in cutoff jeans, a tank top and sandals. "I just thank God we have nice weather, because I'm pretty sure it's still going to rain and that we're still going to get the mudslides."

Bill Patzert, a climatologist at NASA's Jet Propulsion Laboratory in La Cañada Flintridge, said one hypothesis is that El Niño needs to weaken before the storm track can reemerge over Southern California. In 1998, it was a weakening El Niño in January that preceded storms that pounded L.A. in February. Last month, El Niño was still extremely large and potent — about two and a half times the size of the continental United States.

In other words: "This is not too big to fail, but with regard to Southern California, it's too big to succeed," Patzert said.

"I'm still saying: Be patient. In terms of getting Southern California their El Niño fix, this thing has to shrink somewhat.... So if that idea is correct, then we're looking good for March and April."

For the period from Feb. 20 to March 4, the National Weather Service's Climate Prediction Center forecasted a better-than-even chance of above-average rains for Southern California, given the strong El Niño condition in the ocean.

"Even though we haven't seen El Niño pan out" in sending storms to Southern California, said specialist Stuart Seto of the National Weather Service office in Oxnard, "that still doesn't mean we can't see good rains in the latter part of February and in March."

CHAPTER THIRTEEN
WHAT FORCES CAUSE EL NIÑO?

To figure out if a strong El Niño is on the way, forecasters must analyze a series of complex events and processes, ranging from the shifting temperature and structure of subsurface waters across the Pacific, to the evolution of trade winds as well as more localized wind bursts.

One ingredient that appears to be crucial is a multiyear buildup of warm water in the western Pacific. Two types of slow-moving disturbances can also push water upward and downward across the Pacific.Rossby waves travel westward, while Kelvin waves move eastward. In both cases, the sea surface rises, then falls, by up to several inches over a period of weeks to months as the wave passes. Often these effects are transient, but sometimes they can help nudge the Pacific into or out of an El Niño pattern. So can the Madden-Julian Oscillation, a pulse of atmospheric energy generated in the Indian Ocean every few weeks. MJO events can push clusters of showers and thunderstorms eastward across the tropical Pacific.

In addition, to get an El Niño going, something must slow or reverse the trade winds. Once that happens, the displaced wind and water can then reinforce the El Niño state in a feedback process lasting for months. However, it's not easy to turn around those persistent east-to-west winds in the first place. Westerly wind bursts can help do the trick. These clumps of west-to-east wind, pushing directly against the trade winds, can span hundreds of miles and can last a few days to several weeks. They can also kick off the Kelvin waves noted above.

Although La Niña can develop independently of El Niño, it often materializes immediately after an El Niño event, as Kelvin waves and other phenomena cause the ocean to "bounce back," overcorrecting the changes initially brought by El Niño. The La Niña pattern involves a strengthening rather than a partial reversal of both trade winds and the larger Walker Circulation. In part because La Niña more closely resembles the neutral state of the Pacific, it is somewhat easier for a La Niña event to last longer (up to 2–3 years) than an El Niño, which rarely persists for more than a year at a time.

References

1."Independent NASA Satellite Measurements Confirm El Niño is Back and Strong". NASA/JPL.

2. Climate Prediction Center (2005-12-19). "Frequently Asked Questions about El Niño and La Niña". National Centers for Environmental Prediction. Retrieved 2009-07-17.

3. K.E. Trenberth, P.D. Jones, P. Ambenje, R. Bojariu , D. Easterling, A. Klein Tank, D. Parker, F. Rahimzadeh, J.A. Renwick, M. Rusticucci, B. Soden and P. Zhai. "Observations: Surface and Atmospheric Climate Change". In Solomon, S., D. Qin, M. Manning, Z. Chen, M. Marquis, K.B. Averyt, M. Tignor and H.L. Miller. Climate Change 2007: The Physical Science Basis. Contribution of Working Group I to the Fourth Assessment Report of the Intergovernmental Panel on Climate Change. Cambridge, UK: Cambridge University Press. pp. 235–336.

4. "El Niño Information". California Department of Fish and Game, Marine Region.

5 Johnson, Nathaniel C. (19 January 2014)."Atmospheric Science: A boost in big El Niño". Nature Climate Change 4: 90–91. Bibcode:2014NatCC...4...90J.doi:10.1038/nclimate2108 . Retrieved 28 July 2015.(subscription required (help)).

6. Cai, Wenju; Borlace, Simon; Lengaigne, Matthieu; van Rensch, Peter; Collins, Mat; Vecchi, Gabriel; Timmermann, Axel; Santoso, Agus; McPhaden, Michael

J.; Wu, Lixin; England, Matthew H.; Wang, Guojian; Guilyardi, Eric; Jin, Fei-Fei (19 January 2014). "Increasing frequency of extreme El Niño events due to greenhouse warming". Nature Climate Change. pp. 111–116. doi:10.1038/nclimate2100. Retrieved 28 July 2015. (subscription required (help)).

7. Climate Prediction Center (2005-12-19). "ENSO FAQ: How often do El Niño and La Niña typically occur?". National Centers for Environmental Prediction. Retrieved 2009-07-26.

8. National Climatic Data Center (June 2009). "El Niño / Southern Oscillation (ENSO) June 2009". National Oceanic and Atmospheric Administration. Retrieved 2009-07-26.

9. Intergovernmental Panel on Climate Change (2007)."Climate Change 2007: Working Group I: The Physical Science Basis: 3.7 Changes in the Tropics and Subtropics, and in the Monsoons". World Meteorological Organization. Retrieved 2014-07-01.

10. K. Yamasaki, A. Gozolchiani, S. Havlin (2008). "Climate networks around the globe are significantly affected by El Nino" (PDF). Phys. Rev. Lett 100: 228501.arXiv:0709.1792. Bibcode:2008PhRvL.100c8501L.doi:10.1103/PhysRevL ett.100.038501.Sea ocean data assimilation (SODA), 1871–2008". J. Geophys. Res. 116: C02024. Bibcode:2011JGRC..116.2024G.doi:10.1029/2010JC00 6695.

Further reading

• Caviedes, César N. (2001). El Niño in History: Storming Through the Ages. Gainesville: University of Florida Press.ISBN 0-8130-2099-9.

• Fagan, Brian M. (1999). Floods, Famines, and Emperors: El Niño and the Fate of Civilizations. New York: Basic Books.ISBN 0-7126-6478-5.

• Glantz, Michael H. (2001). Currents of change. Cambridge: Cambridge University Press. ISBN 0-521-78672-X.

• Philander, S. George (1990). El Niño, La Niña and the Southern Oscillation. San Diego: Academic Press. ISBN 0-12-553235-0.

• Trenberth, Kevin E. (1997). "The definition of El Niño" (PDF). Bulletin of the American Meteorological Society 78 (12): 2771–7. Bibcode:1997BAMS...78.2771T. doi:10.1175/1520-0477(1997)078<2771:TDOENO>2.0.CO;2. ISSN 1520-0477.

• Kuenzer, C.; Zhao, D.; Scipal, K.; Sabel, D.; Naeimi, V.; Bartalis, Z.; Hasenauer, S.; Mehl, H.; Dech, S.; Waganer, W. (2009). "El Niño southern oscillation influences represented in ERS scatterometer-derived soil moisture data". Applied Geography 29 (4): 463–477. doi:10.1016/j.apgeog.2009.04.004.

• Li, J.; Xie, S.-P.; Cook, E.R.; Morales, M.; Christie, D.; Johnson, N.; Chen, F.; d'Arrigo, R.;

Fowler, A.; Gou, X.; Fang, K. (2013). "El Niño modulations over the past seven centuries". Nature Climate Change 3 (9): 822826.Bibcode:2013NatCC...3..822L.

www.ingramcontent.com/pod-product-compliance
Lightning Source LLC
Chambersburg PA
CBHW071831200526
45169CB00018B/1341